16th Edition IEE Wiring Regulations

Inspection, Testing and Certification

By the same author

Electrical Installation Work

IEE Wiring Regulations: Design and Verification

Wiring Systems and Fault Finding for Installation Electricians

PAT Portable Appliance Testing

Electric Wiring Domestic

IEE Wiring Regulations: Explained and Illustrated

16th Edition IEE Wiring Regulations
Inspection, Testing and Certification

Fifth Edition

Brian Scaddan IEng, MIIE (elec)

AMSTERDAM • BOSTON • HEIDELBERG • LONDON • NEW YORK • OXFORD
PARIS • SAN DIEGO • SAN FRANCISCO • SINGAPORE • SYDNEY • TOKYO

Newnes is an imprint of Elsevier

Newnes
An imprint of Elsevier
Linacre House, Jordan Hill, Oxford OX2 8DP
30 Corporate Drive, Burlington, MA 01803

First published 1996
Second edition 1998
Third edition 2001
Fourth edition 2002
Reprinted 2002, 2003 (twice)
Fifth edition 2005
Reprinted 2005

British Library Cataloguing in Publication Data
A catalogue record for this book is available from the British Library

Library of Congress Cataloguing in Publication Data
A catalogue record for this book is available from the Library of Congress

ISBN 0 7506 65416

For information on all Newnes publications
visit our website at www.newnespress.com

Typeset by Integra Software Services Pvt. Ltd, Pondicherry, India
www.integra-india.com

Printed and bound in Great Britain by Biddles Ltd, King's Lynn, Norfolk

CONTENTS

PREFACE

As a bridge between the 16th Edition course (C&G 2381) and the Design, Erection and Verification course (C&G 2400), the author, in association with the City and Guilds and the NICEIC, developed the Inspection, Testing and Certification course (C&G 2391).

This book was written as an accompaniment to this new 2391 scheme. It is, however, a useful addition to the reference library of contracting electricians and candidates studying for the C&G 2381 and 2400 qualifications.

Brian Scaddan

PREFACE

INTRODUCTION

The IEE Wiring Regulations

Before we embark on the subject of inspection and testing, it is, perhaps, wise to refresh our memories with regards to one or two important topics from the 16th edition (BS 7671 2001).

Clearly the protection of persons and livestock from shock and burns etc. and the prevention of damage to property, are priorities. In consequence, therefore, thorough inspection and testing of an installation and subsequent remedial work where necessary, will significantly reduce the risks.

So, let us start with electric shock, i.e. the passage of current through the body of such magnitude as to have significant harmful effects. Figure 1 illustrates the generally accepted effects of current passing through the human body.

How then are we at risk of electric shock, and how do we protect against it?

There are two ways in which we can be at risk:

1 Touching live parts of equipment or systems that are intended to be live, this is called *direct contact*.
2 Touching conductive parts which are not meant to be live, but have become live due to a fault, this is called *indirect contact*.

The conductive parts associated with indirect contact can either be metalwork of electrical equipment and accessories (Class 1) and that of electrical wiring systems such as conduit and trunking etc. called *exposed conductive parts*, or other metalwork such as pipes, radiators, girders etc. called *extraneous conductive parts*.

1 mA – 2 mA	Barely perceptible, no harmful effects
5 mA – 10 mA	Throw off, painful sensation
10 mA – 15 mA	Muscular contraction, can't let go!
20 mA – 30 mA	Impaired breathing
50 mA and above	Ventricular fibrillation and death

Figure 1 Electric shock

Let us now consider how we may protect against electric shock from whatever source.

Protection against both direct and indirect contact

One method of protecting against shock from both types of contact relies on the fact that the system voltage must not exceed extra low (50 V AC, 120 V ripple free DC), and that all associated wiring etc. is separated from all other circuits of a higher voltage. Such a system is known as a *separated extra low voltage (SELV)*. If a SELV system exceeds 25 V AC, 60 V ripple free DC, then extra protection from *direct contact* must be provided by barriers, enclosures and insulation.

Another method of protection is by the *limitation of discharge of energy* whereby any equipment is so arranged that the current that can flow through the human body or livestock is limited to a safe level. Typical of this method is electric fences.

Protection against direct contact

Apart from SELV, how can we prevent danger to persons and livestock from contact with intentionally live parts? Clearly we must minimize the risk of such contact, and this may be achieved in one or more of the following ways:

1 Insulate any live parts.
2 Ensure that that any uninsulated live parts are housed in suitable enclosures and/or are behind barriers.
3 Place obstacles in the way. (This method would only be used in areas where skilled and/or authorized persons were involved.)
4 Placing live parts out of reach. (Once again, only used in special circumstances, e.g. live rails of overhead travelling cranes.)

The use of a residual current device (RCD) cannot prevent direct contact, but it may be used to supplement any of the other measures taken, provided that it is rated at 30 mA or less and has a tripping time of not more than 40 ms at an operating current of five times its operating current.

It should be noted that RCDs are not the panacea for all electrical ills, they can malfunction, but they are a valid and effective back-up to the other methods. They must not be used as the sole means of protection.

Protection against indirect contact

How can we protect against shock from contact with live, exposed or extraneous conductive parts whilst touching earth, or from contact between live exposed and/or extraneous conductive parts? The most common method is by *earthed equipotential bonding and automatic disconnection of supply (EEBADS).*

All extraneous conductive parts are joined together with a main equipotential bonding conductor and connected to the main earthing terminal, and all exposed conductive parts are connected to the main earthing terminal by the circuit protective conductors. Add to this overcurrent protection that will operate fast enough when a fault occurs and the risk of severe electric shock is significantly reduced.

Other means of protecting against indirect contact may be used, but are less common and some require very strict supervision.

Use of class 2 equipment

Often referred to as double-insulated equipment, this is typical of modern appliances where there is no provision for the connection of a CPC. This does not mean that there should be no exposed conductive parts and that the casing of equipment should be of an insulating material; it simply indicates that live parts are so well insulated that faults from live to conductive parts cannot occur.

Non-conducting location

This is basically an area in which the floor, walls and ceiling are all insulated. Within such an area there must be no protective conductors, and socket outlets will have no earthing connections.

It must not be possible simultaneously to touch two exposed conductive parts, or an exposed conductive part and an extraneous conductive part. This requirement clearly prevents shock current passing through a person in the event of an earth fault, and the insulated construction prevents shock current passing to earth.

Earth-free local equipotential bonding

This is in essence a Faraday cage, where all metal is bonded together but *not* to earth. Obviously, great care must be taken when entering such a zone in order to avoid differences in potential between inside and outside.

The areas mentioned in this and the previous method are very uncommon. Where they do exist, they should be under constant supervision to ensure that no additions or alterations can lessen the protection intended.

Electrical separation

This method relies on a supply from a safety source such as an isolating transformer to BS 3535 which has no earth connection on the secondary side. In the event of a circuit that is supplied from a source developing a live fault to an exposed conductive part, there would be no path for shock current to flow (see Figure 2).

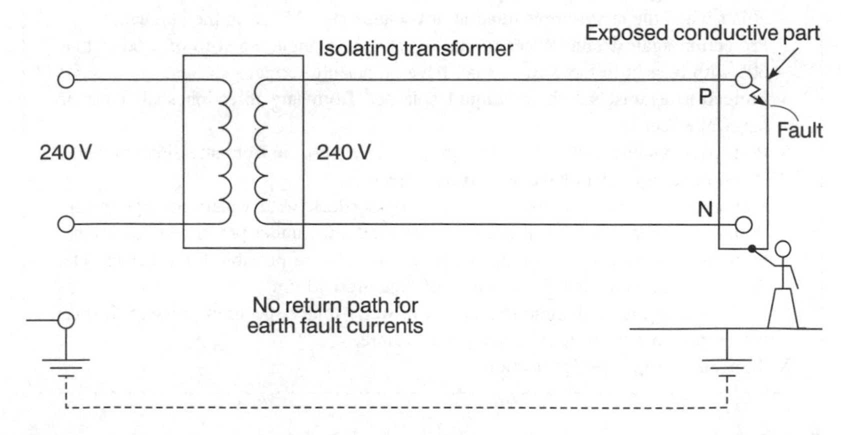

Figure 2

IP codes

First numeral: mechanical protection

0 No protection of persons against contact with live or moving parts inside the enclosure. No protection of equipment against ingress of solid foreign bodies.
1 Protection against accidental or inadvertent contact with live or moving parts inside the enclosure by a large surface of the human body, for example a hand, but not protection against deliberate access to such parts.
2 Protection against ingress of large solid foreign bodies. Protection against contact with live or moving parts inside the enclosure by fingers. Protection against ingress of medium-size solid foreign bodies.
3 Protection against contact with live or moving parts inside the enclosures by tools, wires or such objects of thickness greater than 2.5 mm. Protection against ingress of small foreign bodies.
4 Protection against contact with live or moving parts inside the enclosure by tools, wires or such objects of thickness greater than 1 mm. Protection against ingress of small solid foreign bodies.
5 Complete protection against contact with live or moving parts inside the enclosure. Protection against harmful deposits of dust. The ingress of dust is not totally prevented, but dust cannot enter in an amount sufficient to interfere with satisfactory operation of the equipment enclosed.
6 Complete protection against contact with live or moving parts inside the enclosures. Protection against ingress of dust.

Second numeral: liquid protection

0 No protection.
1 Protection against drops of condensed water. Drops of condensed water falling on the enclosure shall have no harmful effect.
2 Protection against drops of liquid. Drops of falling liquid shall have no harmful effect when the enclosure is tilted at any angle up to 15° from the vertical.
3 Protection against rain. Water falling in rain at an angle equal to or smaller than 60° with respect to the vertical shall have no harmful effect.
4 Protection against splashing. Liquid splashed from any direction shall have no harmful effect.
5 Protection against water jets. Water projected by a nozzle from any direction under stated conditions shall have no harmful effect.
6 Protection against conditions on ships' decks (deck with watertight equipment). Water from heavy seas shall not enter the enclosures under prescribed conditions.
7 Protection against immersion in water. It must not be possible for water to enter the enclosure under stated conditions of pressure and time.
8 Protection against indefinite immersion in water under specified pressure. It must not be possible for water to enter the enclosure.
X Indicates no *specified* protection.

Figure 3

Once again, great care must be taken to maintain the integrity of this type of system, as an inadvertent connection to earth, or interconnection with other circuits, would render the protection useless.

As with direct contact, RCDs are a useful back-up to EEBADS, and their use is essential when hand-held equipment is used outside the main equipotential zone.

The use of enclosures is not limited to protection against direct contact they clearly provide protection against the ingress of foreign bodies and moisture. In order to establish to what degree an enclosure can resist such ingress, reference to the IP code (BSEN 60529) should be made. Figure 3 illustrates part of the IP code.

The most commonly quoted IP codes in the 16th edition are IP2X, IPXXB and IP4X. The X denotes that protection is not specified, not that there is no protection. For example, an enclosure that was to be immersed in water would be classified IPX8, there would be no point using the code IP68.

Note: IPXXB denotes protection against finger contact only.

1
AN OVERVIEW

So, here you are outside the premises, armed with lots of test instruments, a clipboard, a pad of test results sheets, the IEE Regulations, Guidance Notes 3, and an instruction to carry out an inspection and test of the electrical installation therein. Dead easy, you've been told, piece of cake, just poke about a bit, 'Megger' the wiring, write the results down, sign the test certificate and you should be on to the next job within the hour!

Oh! would that it were that simple! What if lethal defects were missed by just 'poking about'? What if other tests should have been carried out which may have revealed serious problems? What if things go wrong after you have signed to say all is in accordance with the Regulations? What if you were not actually competent to carry out the inspection and test in the first place? What if . . . and so on, the list is endless. Inspection, testing and certification is a serious and, in many instances, a complex matter, so let us wind the clock back to the point at which you were about to enter the premises to carry out your tests, and consider the implications of carrying out an inspection and test of an installation.

What are the legal requirements in all of this? Where do you stand if things go wrong? What do you need to do to ensure compliance with the law?

Statutory regulations

The IEE Wiring Regulations (BS 7671) and associated guidance notes are *not* statutory documents, they can however be used in a court of law to prove compliance with statutory requirements such as the Electricity at Work Regulations (EAWR) 1989, which cover all work activity associated with electrical systems in non-domestic situations. A list of other statutory Regulations are given in Appendix 2 of the IEE Regulations. However, it is the Electricity at Work Regulations that are most closely associated with BS 7671, and as such it is worth giving some areas a closer look.

There are thirty-three Regulations in all, twelve of which deal with the special requirements of mines and quarries, and another four with exemptions and extensions outside the UK etc. We are only concerned with the first sixteen Regulations, and Regulation 29, the defence regulation, which we shall come back to later. Let us start then, with a comment on the meaning of *electrical systems and equipment*.

Electrical systems and equipment

According to the EAWR, electrical systems and equipment can encompass anything from power stations to torch or wrist-watch batteries. A battery may not create a shock risk, but may cause burns or injury as a result of attempting to destroy it by fire, whereby explosions may occur. A system can actually include the source of energy, so, a test instrument with its own supply e.g. a continuity tester is a system in itself, and a loop impedance tester, which requires an external supply source, becomes part of the system into which it is connected. From the preceding comments it will be obvious then, that in broad terms, if something is electrical, it is or is part of, an electrical system. So, where does responsibility lie for any involvement with such a system?

The EAWR requires that every employer, employee and self-employed person be responsible for compliance with the Regulations with regards to matters within their control, and as such are known *duty holders*. Where then do you stand as the person about to conduct an inspection and test of an installation? Most certainly, you are a duty holder in that you have control of the installation in

so far as you will ultimately pass the installation as safe or make recommendations to ensure its safety. You also have control of the test instruments which, as already stated are systems in themselves, and control of the installation whilst testing is being carried out.

Any breach of the Regulations may result in prosecution, and unlike the other laws, under this Act you are presumed guilty and have to establish your innocence by invoking the Defence Regulation 29. Perhaps some explanation is needed here. Each of the sixteen Regulations has a status, in that it is either *absolute* or *reasonably practicable*.

Regulations that are *absolute* must be conformed to at all cost, whereas those that are *reasonably practicable* are conformed to, provided that all reasonable steps have been taken to ensure safety. For the contravention of an *absolute* requirement, Regulation 29 is available as a defence in the event of criminal prosecution, provided the accused can demonstrate that they took all reasonable and diligent steps to prevent danger or injury.

No one wants to end up in court accused of negligence, and so we need to be sure that we know what we are doing when we are inspecting and testing.

Apart from the knowledge required competently to carry out the verification process, the person conducting the inspection and test, must be in possession of test instruments appropriate to the duty required of them.

Instruments

In order to fulfil the basic requirements for testing to BS 7671, the following instruments are needed:

1 A low-resistance ohm-meter (continuity tester).
2 An insulation resistance tester.
3 A loop impedance tester.
4 An RCD tester.
5 A prospective fault current (PFC) tester.
6 An approved test lamp or voltage indicator.
7 A proving unit.
8 An earth electrode resistance tester.

Many instrument manufacturers have developed dual or multifunction instruments, hence it is quite common to have continuity and insulation resistance in one unit, loop impedance and PFC in one unit, loop impedance, PFC and RCD tests in one unit etc. However, regardless of the various combinations, let us take a closer look at the individual test instrument requirements.

Low resistance ohmmeters/continuity testers

Bells, buzzers, simple multimeters etc. will all indicate whether or not a circuit is continuous, but will not show the difference between the resistance of say, a 10 m length of 10 mm^2 conductor and a 10 m length of 1 mm^2 conductor. I use this example as an illustration, as it is based on a real experience of testing the continuity of a 10 mm^2 main bonding conductor between gas and water services. The services, some 10 m apart, were at either ends of a domestic premises. The 10 mm^2 conductor, connected to both services, disappeared under the floor, and a measurement between both ends indicated a resistance higher than expected. Further investigation revealed, that just under the floor at each end, the 10 mm^2 conductor had been terminated in a connector block and the join between the two, about 8 m, had been wired with a 1 mm^2 conductor. Only a milli-ohmmeter would have detected such a fault.

> A low resistance ohm-meter should have a no-load source voltage of between 4 V and 24 V, and be capable of delivering an AC or DC short circuit voltage of not less than 200 mA. It should have a resolution (i.e. a detectable difference in resistance) of at least 0.01 milli-ohms.

Insulation resistance testers

An *insulation resistance test* is the correct term for this form of testing, not a '*megger*' test as megger is a manufacturer's trade name, not the name of the test.

An insulation resistance tester must be capable of delivering 1 mA when the required test voltage is applied across the minimum acceptable value of insulation resistance.

Hence, an instrument selected for use on a low voltage (50 V AC–1000 V AC) system should be capable of delivering 1 mA at 500 V across a resistance of 0.5 megohms.

Loop impedance tester

This instrument functions by creating, in effect, an earth fault for a brief moment, and is connected to the circuit via a plug or by 'flying leads' connected separately to phase, neutral, and earth.

The instrument should only allow an earth fault to exist for a maximum of 40 ms, and a resolution of 0.01 ohms is adequate for circuits up to 50 A. Above this circuit rating, the ohmic values become too small to give such accuracy using a standard instrument, and more specialized equipment may be required.

RCD tester

Usually connected by the use of a plug, although 'flying leads' are needed for non-socket outlet circuits, this instrument allows a range of out of balance currents to flow through the RCD to cause its operation within specified time limits.

The test instrument should not be operated for longer than 2 seconds, and it should have a 10 per cent accuracy across the full range of test currents.

PFC tester

Normally one half of a dual, loop impedance/PFC tester, this instrument measures the prospective phase neutral fault current at the point of MEASUREMENT using the same leads as for loop impedance.

Approved test lamp or voltage indicator

A flexible cord with a lamp attached is not an approved device, nor for that matter is the ubiquitous 'testascope' or 'neon screwdriver', which encourages the passage of current, at low voltage, through the body!

A typical approved test lamp is as shown in Figure 1.1.

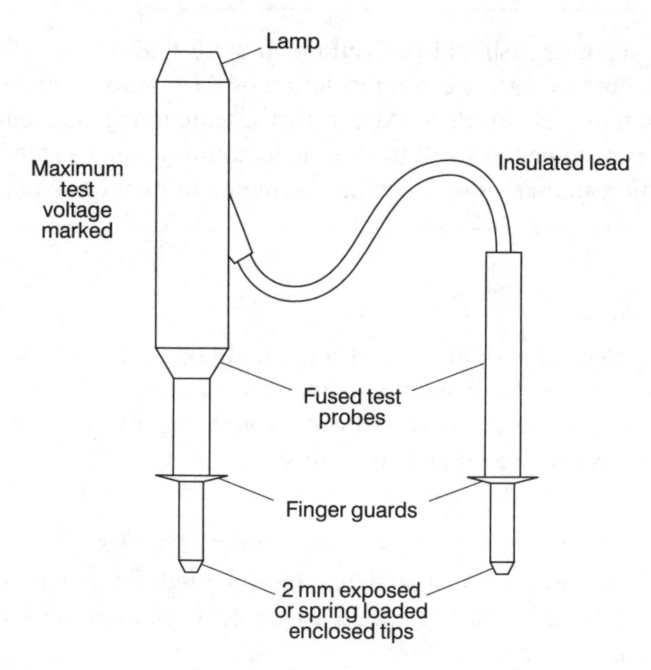

Figure 1.1

The Health and Safety Executive, Guidance Note 38, recommend that the leads and probes associated with test lamps, voltage indicators, voltmeters etc. have the following characteristics:

1 The leads should be adequately insulated and, ideally, fused.
2 The leads should be easily distinguished from each other by colour.
3 The leads should be flexible and sufficiently long for their purpose.
4 The probes should incorporate finger barriers, to prevent accidental contact with live parts.
5 The probes should be insulated and have a maximum of 2 mm of exposed metal, but preferably have spring loaded enclosed tips.

Proving unit

This is an optional item of test equipment, in that test lamps should be proved on a known supply which could, of course, be an adjacent socket or lighting point etc. However, to prove a test lamp on such a known supply may involve entry into enclosures with the associated hazards that such entry could bring. A proving unit is a compact device not much larger than a cigarette packet, which is capable of electronically developing 240 V DC across which the test lamp may be proved. The exception to this are test lamps incorporating 240 V lamps which will not activate from the small power source of the proving unit.

Test lamps must be proved against a voltage similar to that to be tested. Hence, proving test lamps that incorporate an internal check i.e. shorting out the probes to make a buzzer sound, is not acceptable if the voltage to be tested is higher than that delivered by the test lamp.

Care of test instruments

The Electricity at Work Regulations (1989) require that all electrical systems, this includes test instruments, be maintained to prevent danger. This does not restrict such maintenance to just a yearly calibration, but requires equipment to be kept in good condition in order that it is safe to use at all times.

Whilst test instruments and associated leads probes and clips etc., used in the electrical contracting industry are robust in design and manufacture, they still need treating with care and protecting from mechanical damage. Keep test gear in a separate box or case away from tools and sharp objects and always check the general condition of a tester and leads before they are used.

2
INITIAL INSPECTION

Inspection and testing

Circumstances which require an initial verification

New installations or Additions or Alterations

General reasons for initial verification

1 To ensure equipment and accessories are to a relevant standard.
2 To prove compliance with BS 7671.
3 To ensure that the installation is not damaged so as to impair safety.

Information required

Assessment of general characteristics sections 311, 312 and 313 together with information such as drawings, charts etc. in accordance with Reg. 514–09–01.

Documentation required and to be completed

Electrical Installation Certificate signed or authenticated for the design and construction and then for the inspection and test (could be the same person). A schedule of test results and an inspection schedule must accompany an Electrical Installation Certificate.

Sequence of tests

1 Continuity of all protective conductors.
2 Continuity of ring final circuit conductors.
3 Insulation resistance.
4 Site applied insulation.
5 Protection by separation of circuits.
6 Protection against direct contact by barriers and enclosures provided during erection.
7 Insulation of non-conducting floors and walls.
8 Polarity.
9 Earth electrode resistance.
10 Earth fault loop impedance.
11 Prospective fault current.
12 Functional testing.

Before any testing is carried out, a detailed physical inspection must be made to ensure that all equipment is to a relevant British or Harmonized European Standard, and that it is erected/installed in compliance with the IEE Regulations, and that it is not damaged such that it could cause danger. In order to comply with these requirements, the Regulations give a check list of some eighteen items that, where relevant, should be inspected.

However, before such an inspection, and test for that matter, is carried out, certain information *must* be available to the verifier. This information is the result of the Assessment of General Characteristics required by IEE Regulations Part 3, Sections 311, 312 and 313, and drawings, charts, and similar information relating to the installation. It is at this point that most readers who work in the real world of electrical installation will be lying on the floor laughing hysterically.

Let us assume that the designer and installer of the installation are competent professionals, and all of the required documentation is available.

Interestingly, one of the items on the check list *is* the presence of diagrams, instructions and similar information. If these are missing then there is a departure from the Regulations.

Another item on the list is the verification of conductors for current carrying capacity and voltage drop in accordance with the design. How on earth can this be verified without all the information? A 30 A Type B circuit breaker (CB) or type 2 MCB protecting a length of 4 mm^2 conductor may look reasonable, but is it correct, and are you prepared to sign to say that it is unless you are sure? Let us look then at the general content of the check list.

1 **Connection of conductors** Are terminations electrically and mechanically sound, is insulation and sheathing removed only to a minimum to allow satisfactory termination?
2 **Identification of conductors** Are conductors correctly identified in accordance with the Regulations?
3 **Routing of cables** Are cables installed such that account is taken of external influences such as mechanical damage, corrosion, heat etc?
4 **Conductor selection** Are conductors selected for current carrying capacity and voltage drop in accordance with the design?
5 **Connection of single pole devices** Are single pole protective and switching devices connected in the phase conductor only?
6 **Accessories and equipment** Are all accessories and items of equipment correctly connected?
7 **Thermal effects** Are fire barriers present where required and protection against thermal effects provided?
8 **Protection against shock** What methods have been used to provide protection against direct and indirect contact?
9 **Mutual detrimental influence** Are wiring systems installed such that they can have no harmful effect on non-electrical systems, or that systems of different currents or voltages are segregated where necessary?
10 **Isolation and switching** Are there appropriate devices for isolation and switching correctly located and installed?
11 **Undervoltage** Where undervoltage may give rise for concern, are there protective devices present?
12 **Protective devices** Are protective and monitoring devices correctly chosen and set to ensure protection against indirect contact and/or overcurrent?

13 **Labelling** Are all protective devices, switches (where necessary) and terminals correctly labelled?
14 **External influences** Have all items of equipment and protective measures been selected in accordance with the appropriate external influences?
15 **Access** Are all means of access to switchgear and equipment adequate?
16 **Notices and signs** Are danger notices and warning signs present?
17 **Diagrams** Are diagrams, instructions and similar information relating to the installation available?
18 **Erection methods** Have all wiring systems, accessories and equipment been selected and installed in accordance with the requirements of the Regulations, and are fixings for equipment adequate for the environment?

So, we have now inspected all relevant items, and provided that there are no defects that may lead to a dangerous situation when testing, we can now start the actual testing procedure.

3
TESTING CONTINUITY OF PROTECTIVE CONDUCTORS

All protective conductors, including main equipotential and supplementary bonding conductors must be tested for continuity using a low resistance ohmmeter.

For main equipotential bonding there is no single fixed value of resistance above which the conductor would be deemed unsuitable. Each measured value, if indeed it is measurable for very short lengths, should be compared with the relevant value for a particular conductor length and size. Such values are shown in Table 3.1.

Table 3.1 Resistance (Ω) of copper conductors at 20°C

CSA	*Length (metres)*									
(mm^2)	5	10	15	20	25	30	35	40	45	50
1.0	0.09	0.18	0.27	0.36	0.45	0.54	0.63	0.72	0.82	0.9
1.5	0.06	0.12	0.18	0.24	0.3	0.36	0.43	0.48	0.55	0.6
2.5	0.04	0.07	0.11	0.15	0.19	0.22	0.26	0.03	0.33	0.37
4.0	0.023	0.05	0.07	0.09	0.12	0.14	0.16	0.18	0.21	0.23
6.0	0.02	0.03	0.05	0.06	0.08	0.09	0.11	0.13	0.14	0.16
10.0	0.01	0.02	0.03	0.04	0.05	0.06	0.063	0.07	0.08	0.09
16.0	0.006	0.01	0.02	0.023	0.03	0.034	0.04	0.05	0.05	0.06
25.0	0.004	0.007	0.01	0.015	0.02	0.022	0.026	0.03	0.033	0.04
35.0	0.003	0.005	0.008	0.01	0.013	0.016	0.019	0.02	0.024	0.03

Where a supplementary equipotential bonding conductor has been installed between *simultaneously accessible* exposed and extraneous conductive parts, because circuit disconnection times cannot be met, then the resistance (R) of the conductor, must be equal to or less than 50/Ia. In the case of construction sites and agricultural or horticultural installations the 50 is replaced by 25.

So, $R \leqslant 50/\text{Ia}$, or $25/\text{Ia}$, where 50, and 25, are the voltages above which exposed metalwork should not rise, and Ia is the minimum current causing operation of the circuit protective device within 5 secs.

For example, suppose a 45 A BS 3036 fuse protects a cooker circuit, the disconnection time for the circuit cannot be met, and so a supplementary bonding conductor has been installed between the cooker case and an adjacent central heating radiator. The resistance (R) of that conductor should not be greater than 50/Ia, and Ia in this case is 145 A (see Figure 3.2B of the IEE Regulations)

i.e. $50/145 = 0.34\,\Omega$

How then, do we conduct a test to establish continuity of main or supplementary bonding conductors? Quite simple really, just connect the leads from a low resistance ohmmeter to the ends of the bonding conductor (Figure 3.1). One end should be disconnected from its bonding clamp, otherwise any measurement may

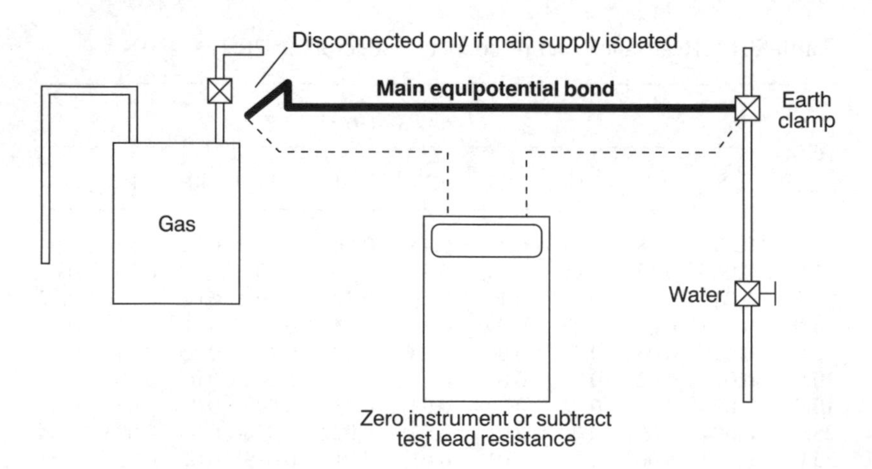

Figure 3.1

include the resistance of parallel paths of other earthed metalwork. Remember to zero the instrument first or, if this facility is not available, record the resistance of the test leads so that this value can be subtracted from the test reading.

IMPORTANT NOTE: If the installation is in operation, then *never* disconnect main bonding conductors unless the supply can be isolated. Without isolation, persons and livestock are at risk of electric shock.

The continuity of circuit protective conductors may be established in the same way, but a second method is preferred, as the results of this second test indicate the value of $(R_1 + R_2)$ for the circuit in question.

The test is conducted in the following manner:

1 Temporarily link together the phase conductor and cpc of the circuit concerned in the distribution board or consumer unit.
2 Test between phase and cpc at EACH outlet in the circuit. A reading indicates continuity.
3 Record the test result obtained at the furthest point in the circuit. This value is $(R_1 + R_2)$ for the circuit.

Figure 3.2 illustrates the above method.

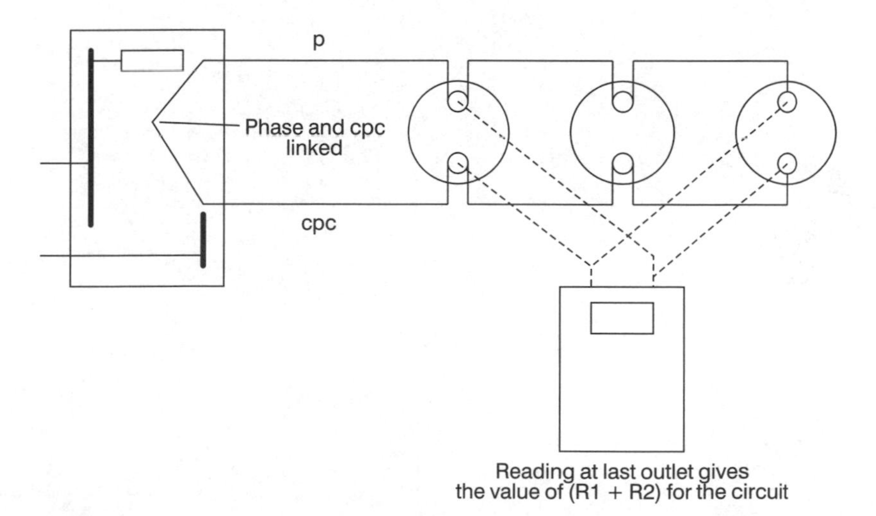

Figure 3.2

There may be some difficulty in determining the $(R_1 + R_2)$ values of circuits in installations that comprise steel conduit and trunking, and/or SWA and mims cables because of the parallel earth paths that are likely to exist. In these cases, continuity tests may have to be carried out at the installation stage before accessories are connected or terminations made off as well as after completion.

Although it is no longer considered good working practice to use steel conduit or trunking as a protective conductor, it is permitted, and hence its continuity must be proved. The enclosure must be inspected along its length to ensure that it is sound and then the standard low resistance test is performed.

4 TESTING CONTINUITY OF RING FINAL CIRCUIT CONDUCTORS

There are two main reasons for conducting this test:

1 To establish that interconnections in the ring do not exist.
2 To ensure that the cpc is continuous, and indicate the value of $(R_1 + R_2)$ for the ring.

What then are interconnections in a ring circuit, and why is it important to locate them? Figure 4.1 shows a ring final circuit with an interconnection.

The most likely cause of the situation shown in Figure 4.1 is where a DIY enthusiast has added sockets P, Q, R and S to an existing ring A, B, C, D, E and F.

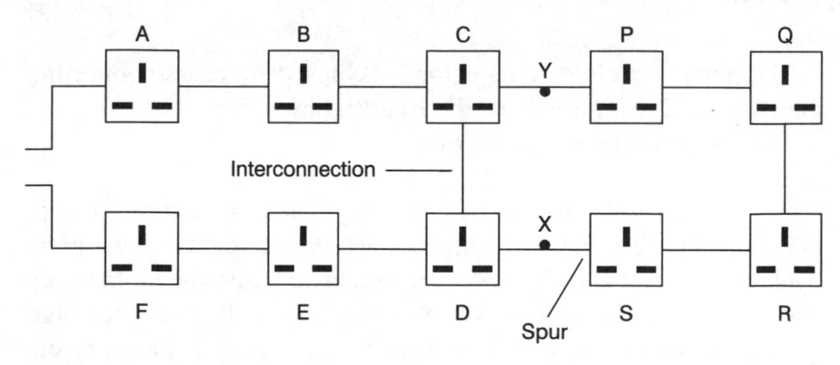

Figure 4.1

In itself there is nothing wrong with this. The problem arises if a break occurs at, say, point Y, or the terminations fail in socket C or P. Then there would be four sockets all fed from the point X which would then become a spur.

So, how do we identify such a situation with or without breaks at point 'Y'? A simple resistance test between the ends of the phase, neutral or circuit protective conductors will only indicate that a circuit exists, whether there are interconnections or not. The following test method is based on the philosophy that the resistance measured across any diameter of a perfect circle of conductor will always be the same value (Figure 4.2).

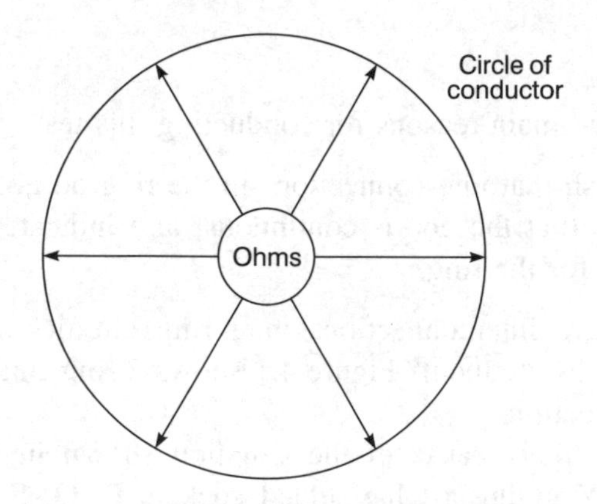

Figure 4.2

The perfect circle of conductor is achieved by cross-connecting the phase and neutral legs of the ring (Figure 4.3).

The test procedure is as follows:

1 Identify the opposite legs of the ring. This is quite easy with sheathed cables, but with singles, each conductor will have to be identified, probably by taking resistance measurements between each one and the closest socket outlet. This will give three high readings and three low readings thus establishing the opposite legs.

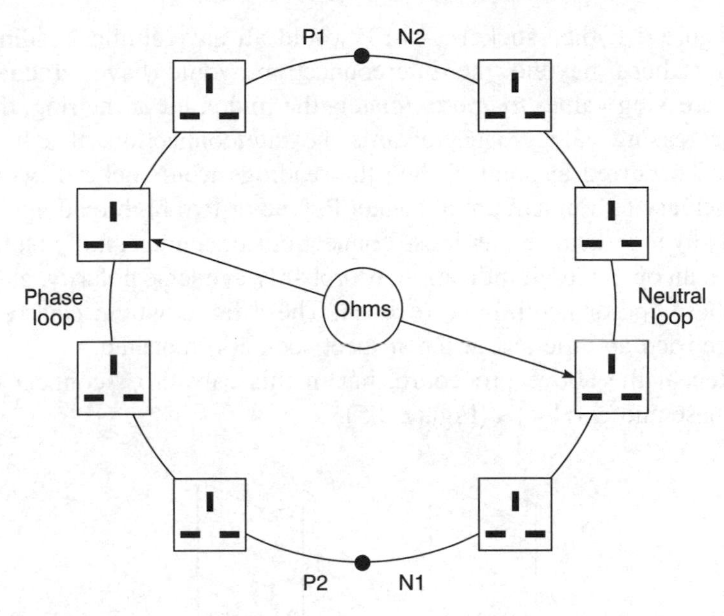

Figure 4.3

2 Take a resistance measurement between the ends of each conductor loop. Record this value.
3 Cross-connect the opposite ends of the phase and neutral loops (Figure 4.4).
4 Measure between phase and neutral at each socket on the ring. The readings obtained should be, for a perfect ring, substantially the same. If an interconnection existed such as shown in

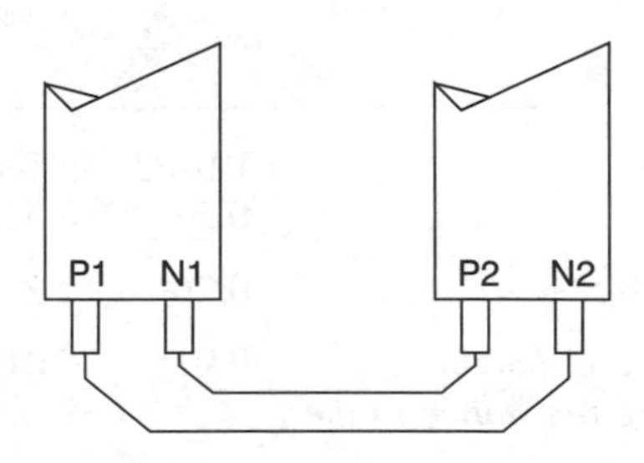

Figure 4.4

Figure 4.1, then sockets A to F would all have similar readings, and those beyond the interconnection would have gradually increasing values to approximately the mid point of the ring, then decreasing values back towards the interconnection. If a break had occurred at point Y then the readings from socket S would increase to a maximum at socket P. One or two high readings are likely to indicate either loose connections or spurs. A null reading, i.e. an open circuit indication, is probably a reverse polarity, either phase-cpc or neutral-cpc reversal. These faults would clearly be rectified and the test at the suspect socket(s) repeated.

5 Repeat the above procedure, but in this case cross-connect the phase and cpc loops (Figure 4.5).

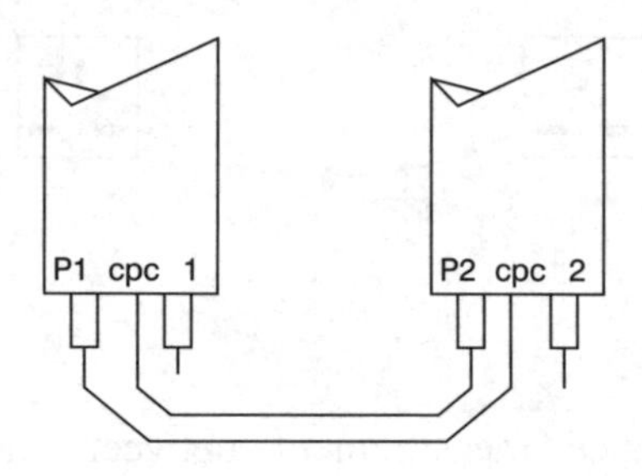

Figure 4.5

In this instance, if the cable is of the flat twin type, the readings at each socket will increase very slightly and then decrease around the ring. This difference, due to the phase and cpc being different sizes, will not be significant enough to cause any concern. The measured value is very important, it is $R_1 + R_2$ for the ring.

Table 4.1

	P1–P2	N1–N2	cpc 1 – cpc 2
Initial measurements	0.52	0.52	0.86
Reading at each socket	0.26	0.26	0.32–0.34
For spurs, each metre in length will add the following resistance to the above values	0.015	0.015	0.02

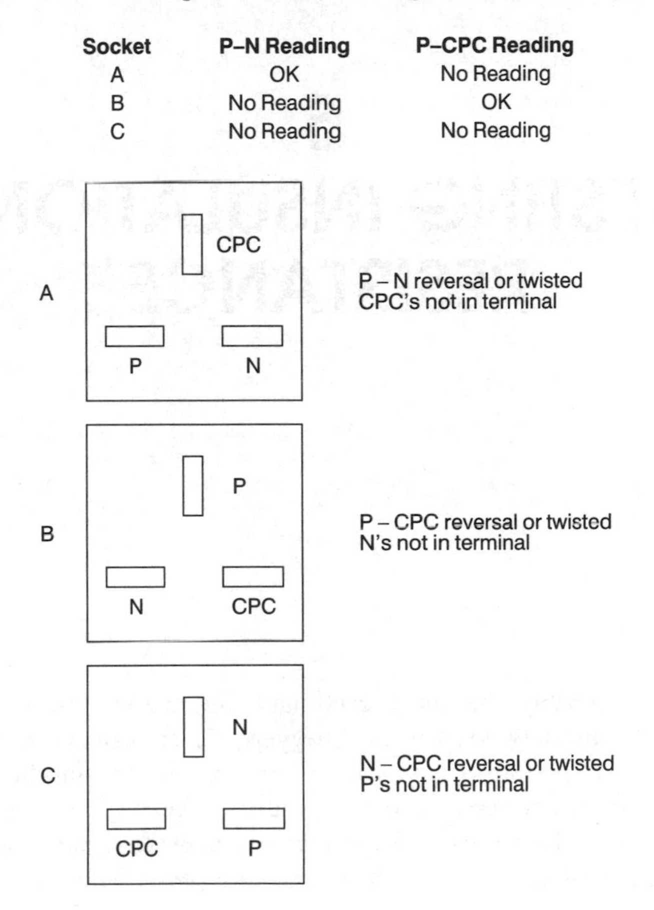

Socket	P–N Reading	P–CPC Reading
A	OK	No Reading
B	No Reading	OK
C	No Reading	No Reading

Figure 4.6

As before, loose connections, spurs and, in this case, P–N cross-polarity, will be picked up.

The details in Table 4.1 are typical approximate ohmic values for a healthy 70 m ring final circuit wired in 2.5/1.5 flat twin and cpc cable. (In this case the CPC will be approximately 1.67× the P or N resistance.)

As already mentioned null readings may indicate a reverse polarity. They could also indicate twisted conductors not in their terminal housing. The examples shown in Figure 4.6 may help to explain these situations.

5
TESTING INSULATION RESISTANCE

This is probably the most used and yet abused test of them all. Affectionately known as 'meggering', an *insulation resistance test* is performed in order to ensure that the insulation of conductors, accessories and equipment is in a healthy condition, and will prevent dangerous leakage currents between conductors and between conductors and earth. It also indicates whether any short circuits exist.

Insulation resistance, as just discussed, is the resistance measured between conductors and is made up of countless millions of resistances in parallel (Figure 5.1).

The more resistances there are in parallel, the *lower* the overall resistance, and in consequence, the longer a cable the lower the insulation resistance. Add to this the fact that almost all installation circuits are also wired in parallel, it becomes apparent that tests on large installations may give, if measured as a whole, pessimistically low values, even if there are no faults.

Under these circumstances, it is usual to break down such large installations into smaller sections, floor by floor, sub-main by

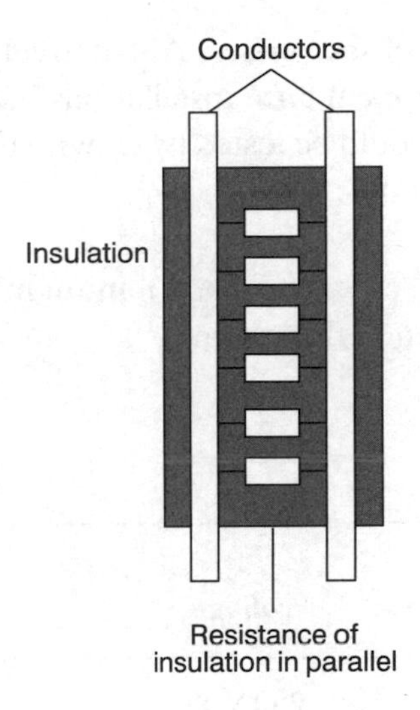

Figure 5.1

submain etc. This also helps, in the case of periodic testing, to minimize disruption. The test procedure is as follows:

1 Disconnect all items of equipment such as capacitors and indicator lamps as these are likely to give misleading results. Remove any items of equipment likely to be damaged by the test, such as dimmer switches, electronic timers etc. Remove all lamps and accessories and disconnect fluorescent and discharge fittings. Ensure that the installation is disconnected from the supply, all fuses are in place, and MCBs and switches are in the on position. In some instances it may be impracticable to remove lamps etc. and in this case the local switch controlling such equipment may be left in the off position.
2 Join together all live conductors of the supply and test between this join and earth. Alternatively, test between each live conductor and earth in turn.
3 Test between phase and neutral. For three phase systems, join together all phases and test between this join and neutral. Then

test between each of the phases. Alternatively, test between each of the live conductors in turn. Installations incorporating two-way lighting systems should be tested twice with the two-way switches in alternative positions.

Table 5.1 gives the test voltages and minimum values of insulation resistance for ELV and LV systems.

Table 5.1

System	*Test voltage*	*Minimum insulation resistance*
SELV and PELV	250 V DC	0.25 MΩ
LV up to 500 V	500 V DC	0.5 MΩ
Over 500 V	1000 V DC	1.0 MΩ

If a value of less than 2 MΩ is recorded it may indicate a situation where a fault is developing, but as yet still complies with the minimum permissible value. In this case each circuit should be tested separately in order to locate the problem.

Example

An installation comprising six circuits have individual insulation resistances of 2.5 MΩ, 8 MΩ, 200 MΩ, 200 MΩ, 200 MΩ and 200 MΩ, and so the total insulation resistance will be:

$$\frac{1}{R_t} = \frac{1}{2.5} + \frac{1}{8} + \frac{1}{200} + \frac{1}{200} + \frac{1}{200} + \frac{1}{200}$$

$$= 0.4 + 0.125 + 0.005 + 0.005 + 0.005 + 0.005$$

$$= 0.545$$

$$R_t = \frac{1}{0.545}$$

$$= 1.83\text{M}\,\Omega$$

This is clearly greater than the 0.5 MΩ minimum but less than 2 MΩ, but as all circuits are greater than 2 MΩ the system could be considered satisfactory.

6
SPECIAL TESTS

The next four tests are special in that they are not often required in the general type of installation. They also require special test equipment. In consequence, the requirements for these tests will only be briefly outlined in this chapter.

Site applied insulation

When insulation is applied to live parts during the erection process on site in order to provide protection against direct contact, then a test has to be performed to show that the insulation can withstand a high voltage equivalent to that specified in the BS for similar factory built equipment.

If supplementary insulation is applied to equipment on site, to provide protection against indirect contact, then the voltage withstand test must be applied, and the insulating enclosure must afford a degree of protection of not less than IP2X or IPXXB.

Protection by separation of circuits

When SELV or PELV is used as a protective measure, then the separation from circuits of a higher voltage has to be verified by an insulation resistance test at a test voltage of 250 V and result in a minimum insulation resistance of 0.25 MΩ. If the circuit is at low voltage and supplied from, say, a BS 3535 transformer the test is at 500 V with a minimum value of 0.555 MΩ.

Protection by barriers or enclosures

If, on site, protection against direct contact is provided by fabricating an enclosure or erecting a barrier, it must be shown that the enclosure can provide a degree of protection of at least IP2X or IPXXB. Readily accessible horizontal top surfaces only should be to at least IP4X.

An enclosure having a degree of protection IP2X can withstand the ingress of fingers and solid objects exceeding 12 mm diameter. IPXXB is protection against finger contact only. IP4X gives protection against wires and solid objects exceeding 1 mm in diameter.

The test for IP2X or IPXXB is conducted with a 'standard test finger' which is supplied at a test voltage not less than 40 V and no more than 50 V. One end of the finger is connected in series with a lamp and live parts in the enclosure. When the end of the finger is introduced into the enclosure, provided the lamp does not light then the protection is satisfactory.

The test for IP4X is conducted with a rigid 1 mm diameter wire with its end bent at right angles. Protection is afforded if the wire does not enter the enclosure.

Protection by non-conducting location

This is a rare location and demands specialist equipment to measure the insulation resistance between insulated floors and walls at various points.

7
TESTING POLARITY

This simple test, often overlooked, is just as important as all the others, and many serious injuries and electrocutions could have been prevented if only polarity checks had been carried out.

The requirements are:

1 All fuses and single pole switches are in the phase conductor.
2 The centre contact of an Edison screw type lampholder is connected to the phase conductor (except E14 & 27 types).
3 All socket outlets and similar accessories are correctly wired.

Although polarity is towards the end of the recommended test sequence, it would seem sensible, on lighting circuits, for example, to conduct this test at the same time as that for continuity of CPCs (Figure 7.1).

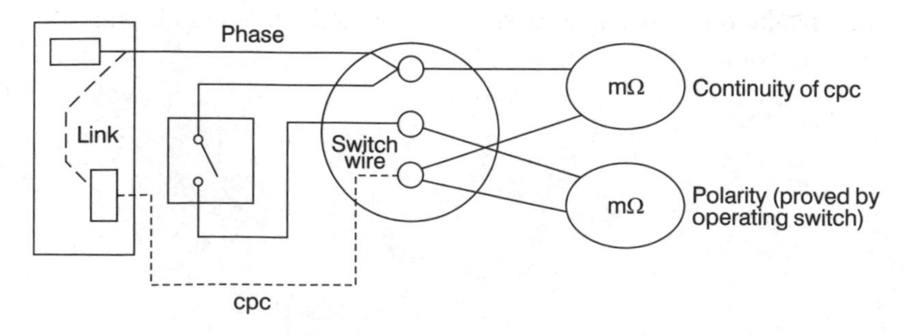

Figure 7.1

As discussed earlier, polarity on ring final circuit conductors, is achieved simply by conducting the ring circuit test. For radial socket outlet circuits, however, this is a little more difficult. The continuity of the CPC will have already been proved by linking phase and CPC and measuring between the same terminals at each socket. Whilst a phase-CPC reversal would not have shown, a phase-neutral reversal would, as there would have been no reading at the socket in question. This would have been remedied, and so only phase-CPC reversals need to be checked. This can be done by linking together phase and neutral at the origin and testing between the same terminals at each socket. A phase-CPC reversal will result in no reading at the socket in question.

When the supply is connected, it is important to check that the incoming supply is correct. This is done using an approved voltage indicator at the intake position or close to it.

8 TESTING EARTH FAULT LOOP IMPEDANCE

This is very important but sadly, poorly understood. So let us remind ourselves of the component parts of the earth fault loop path (Figure 8.1). Starting at the point of fault:

1 The CPC.
2 The earthing conductor and earthing terminal.
3 The return path via the earth for TT systems, and the metallic return path in the case of TN–S or TN–C–S systems. In the latter case the metallic return is the PEN conductor.
4 The earthed neutral of the supply transformer.
5 The transformer winding.
6 The phase conductor back to the point of fault.

Overcurrent protective devices must, under earth fault conditions, disconnect fast enough to reduce the risk of electric shock. This is achieved if the actual value of the earth fault loop impedance does not exceed the tabulated maximum values given in the IEE regulations.

The purpose of the test, therefore, is to determine the actual value of the loop impedance (Z_s), for comparison with those maximum values, and it is conducted as follows:

1 Ensure that all main equipotential bonding is in place.
2 Connect the test instrument either by its BS 4363 plug, or the ‘flying leads’, to the phase, neutral and earth terminals at

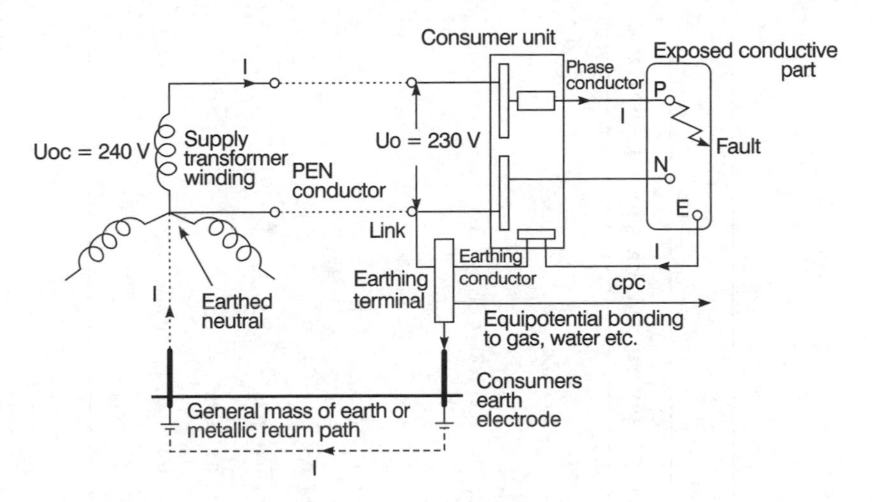

Figure 8.1

the remote end of the circuit under test. (If a neutral is not available, e.g. in the case of a three phase motor, connect the neutral probe to earth.)

3 Press to test and record the value indicated.

It must be understood, that this instrument reading is *not valid for direct comparison with the tabulated maximum values*, as account must be taken of the ambient temperature at the time of test, and the maximum conductor operating temperature, both of which will have an effect on conductor resistance. Hence, the $(R_1 + R_2)$ could be greater at the time of fault than at the time of test.

So, our measured value of Z_s must be corrected to allow for these possible increases in temperature occurring at a later date. This requires actually measuring the ambient temperature and applying factors in a formula.

Clearly this method of correcting Z_s is time consuming and unlikely to be commonly used. Hence, a rule of thumb method may be applied which simply requires that the measured value of Z_s does not exceed 3/4 of the appropriate tabulated value. Table 8.3 gives the 3/4 values of tabulated loop impedance for direct comparison with measured values.

Table 8.1 Values of loop impedance for comparison with test readings

Protection	*Disconnection time (s)*		*Rating of protection*																		
			5A	6A	10A	15A	16A	20A	25A	30A	32A	40A	45A	50A	60A	63A	80A	100A	125A	160A	200A
BS 3036 fuse	0.4	Z_s max	7.5	–	–	2	–	1.38	–	0.85	–	–	0.46								
	5	Z_s max	13.9	–	–	4.1	–	3	–	2.97	–	–	1.25	–	0.87	–	–	0.42			
BS 88 fuse	0.4	Z_s max	–	6.66	4	–	2.11	1.38	1.12	–	0.82	0.64	–	0.47							
	5	Z_s max	–	10.5	5.8	–	3.27	2.28	1.8	–	1.44	1.05	–	0.82	–	0.64	0.45	0.33	0.26	0.2	0.14
BS 1361 fuse	0.4	Z_s max	8.17	–	–	2.57	–	1.33	–	0.9	–	–	0.45								
	5	Z_s max	12.8	–	–	3.9	–	2.19	–	1.44	–	–	0.75	0.54	–	0.39	0.28				
BS 3871 MCB Type 1	0.4&5	Z_s max	9	7.5	4.5	3	2.81	2.25	1.8	1.5	1.41	1.12	1	0.9	–	0.71					
BS 3871 MCB Type 2	0.4&5	Z_s max	5.14	4.28	2.57	1.71	1.6	1.28	1.02	0.85	0.8	0.64	0.57	0.51	–	0.4					
BS 3871 MCB Type 3	0.4&5	Z_s max	3.6	3	1.8	1.2	1.12	0.9	0.72	0.6	0.56	0.45	0.4	0.36	–	0.28					
BS EN 60898 CB Type B	0.4&5	Z_s max	–	6	3.6	–	2.25	1.8	1.44	–	1.12	0.9	0.8	0.72	–	0.57					
BS EN 60898 CB Type C	0.4&5	Z_s max	3.6	3	1.8	1.2	1.12	0.9	0.72	0.6	0.56	0.45	0.4	0.36	–	0.28					
BS EN 60898 CB Type D	0.4&5	Z_s max	1.8	1.5	0.9	0.6	0.56	0.45	0.36	0.3	0.28	0.22	0.2	0.18	–	0.14					

In effect, a loop impedance test places a phase/earth fault on the installation, and if an RCD is present it may not be possible to conduct the test as the device will trip out each time the loop impedance tester button is pressed. Unless the instrument is of a type that has built-in guard against such tripping, the value of Z_s will have to be determined from measured values of Z_e and $(R_1 + R_2)$, and the 3/4 rule applied.

IMPORTANT NOTE: Never short out an RCD in order to conduct this test.

As a loop impedance test creates a high earth fault current, albeit for a short space of time, some lower rated MCBs may operate resulting in the same situation as with an RCD, and Z_s will have to be calculated. It is not really good practice temporarily to replace the MCB with one of a higher rating.

External loop impedance Z_e

The value of Z_e is measured at the intake position on the supply side and with the means of earthing disconnected. Unless the installation can be isolated from the supply, this test should not be carried out, as a potential shock risk will exist with the supply on and the earthing disconnected.

Prospective fault current

This would normally be carried out at the same time as the measurement for Z_e using a PFC or PSCC tester. If this value cannot be measured it must be ascertained by either enquiry or calculation.

9 TESTING EARTH ELECTRODE RESISTANCE

In many rural areas, the supply system is TT and hence reliance is placed on the general mass of earth for a return path under earth fault conditions. Connection to earth is made by an electrode, usually of the rod type, and preferably installed as shown in Figure 9.1.

In order to determine the resistance of the earth return path, it is necessary to measure the resistance that the electrode has with earth. If we were to make such measurements at increasingly longer distances from the electrode, we would notice an increase in

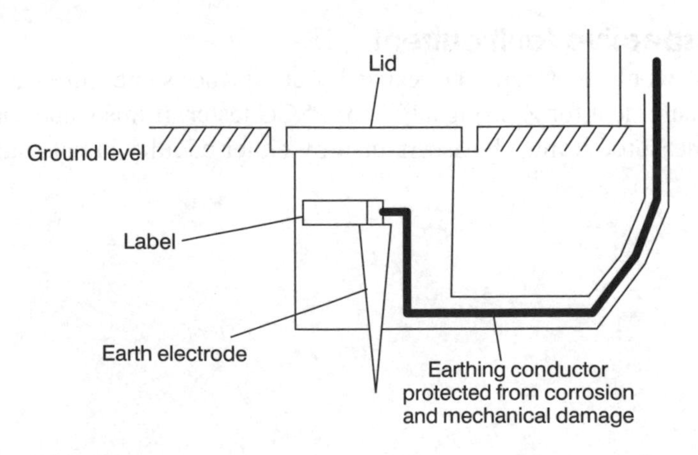

Figure 9.1

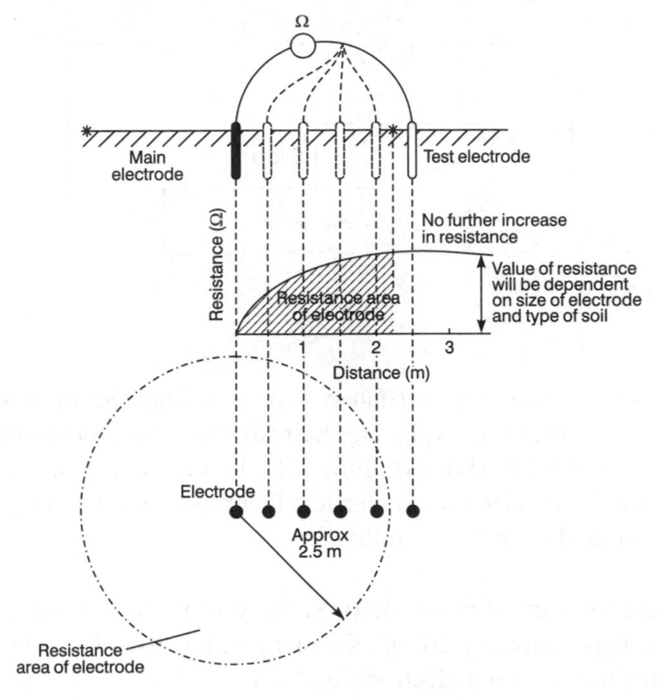

Figure 9.2

resistance up to about 2.5–3 m from the rod, after which no further increase in resistance would be noticed (Figure 9.2).

The maximum resistance recorded is the electrode resistance and the area that extends the 2.5–3 m beyond the electrode is known as the earth electrode resistance area.

There are two methods of making the measurement, one using a proprietary instrument, and the other using a loop impedance tester.

Method 1 – protection by overcurrent device

This method is based on the principle of the potential divider (Figure 9.3).

By varying the position of the slider the resistance at any point may be calculated from $R = V/I$.

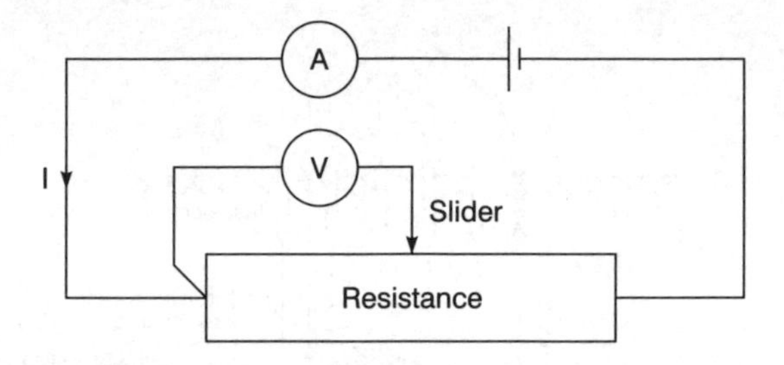

Figure 9.3

The earth electrode resistance test is conducted in a similar fashion, with the earth replacing the resistance and a potential electrode replacing the slider (Figure 9.4). In Figure 9.4 the earthing conductor to the electrode under test is temporarily disconnected.

The method of test is as follows:

1 Place the current electrode (C2) away from the electrode under test, approximately 10 times its length, i.e. 30 m for a 3 m rod.
2 Place the potential electrode mid way.
3 Connect test instrument as shown.
4 Record resistance value.

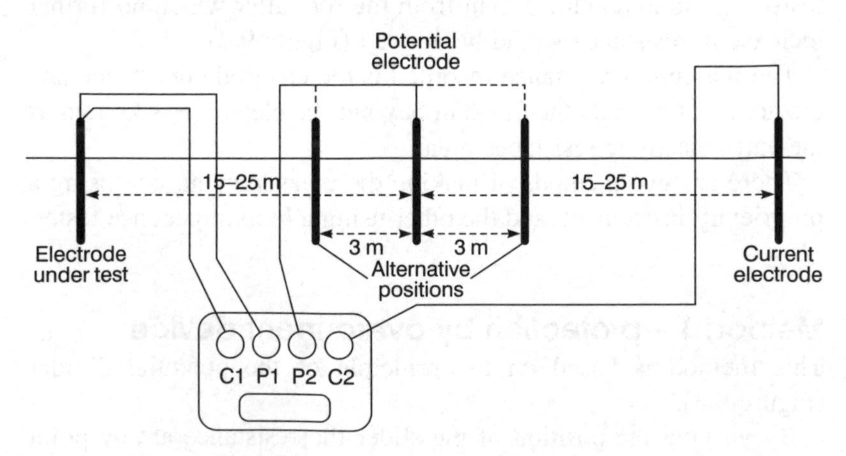

Figure 9.4

5 Move the potential electrode approximately 3 m either side of the mid position, and record these two readings.
6 Take an average of these three readings (this is the earth electrode resistance).
7 Determine the maximum deviation or difference of this average from the three readings.
8 Express this deviation as a percentage of the average reading.
9 Multiply this percentage deviation by 1.2.
10 Provided this value does not exceed a figure of 5% then the accuracy of the measurement is considered acceptable.

If three readings obtained from an earth electrode resistance test were 181 Ω, 185 Ω and 179 Ω. What is the value of the electrode resistance and is the accuracy of the measurement acceptable?

$$\text{Average value} = \frac{181 + 185 + 179}{3}$$

$$= 181.67\,\Omega$$

$$\text{Maximum deviation} = 185 - 181.67$$

$$= 3.33$$

$$\text{Expressed as a percentage of the average} = \frac{3.33 \times 100}{181.67}$$

$$= 1.83\%$$

$$\text{Measurement accuracy} = 1.83\% \times 1.2 = 2.2\%$$

(which is acceptable)

For TT systems the result of this test will indicate compliance if the product of the electrode resistance and the operating current of the overcurrent device does not exceed 50 V.

Method 2 – protection by a residual current device

In this case, an earth fault loop impedance test is carried out between the incoming phase terminal and the electrode (a standard test for Z_e).

The value obtained is added to the cpc resistance of the protected circuits and this value is multiplied by the operating current of the RCD. The resulting value should not exceed 50 V. If it does, then Method 1 should be used to check the actual value of the electrode resistance.

10
FUNCTIONAL TESTING

RCD RCBO operation

Where RCDs RCBOs are fitted, it is essential that they operate within set parameters. The RCD testers used are designed to do just this, and the basic tests required are as follows:

1 Set the test instrument to the rating of the RCD.
2 Set the test instrument to half rated trip $\left(\frac{1}{2}I_{\Delta n}\right)$.
3 Operate the instrument and the RCD should not trip.
4 Set the instrument to deliver the full rated tripping current of the RCD ($I_{\Delta n}$).
5 Operate the instrument and the RCD should trip out in the required time.
6 For 30 mA RCD: or less, operate at $5 \times I_{\Delta n}$, should trip in 40 ms.

Table 10.1

RCD type	*½ rated*	*Full trip current*
BS 4239 and BS 7288 sockets	No trip	<200 ms
BS 4239 with time delay	No trip	$\frac{1}{2}$ time delay + 200 ms-time delay + 200 ms
BS EN 61009 or BS EN 61009 rcbo	No trip	<300 ms
As above but Type S with time delay	No trip	>130 to <500 ms

There seems to be a popular misconception regarding the ratings and uses of RCD's in that they are the panacea for all electrical ills and the only useful rating is 30 mA!

Firstly, RCD's are not fail safe devices, they are electromechanical in operation and can malfunction. Secondly, general purpose RCD's are manufactured in ratings from 5 mA to 100 mA and have many uses. Let us first deal with RCD's rated at 30 mA or less. The accepted lethal level of shock current is 50 mA and hence RCD's rated at 30 mA or less would be appropriate for use where shock is an increased risk. BS 7671 indicates that RCD's of 30 mA or less should be used in the following situations:

1 To protect circuits supplying hand held equipment outside the equipotential zone.
2 To protect all socket outlet circuits in a TT system installation.
3 To protect all socket outlets in a caravan park.
4 To provide supplementary protection against Direct contact.
5 For fixed current using equipment in bathrooms.

In all these cases and apart from conducting the tests already mentioned, it is required that the RCD be injected with a current five times its operating current and the tripping time should not exceed 40 ms.

Where loop impedence values cannot be met, RCD's of an appropriate rating can be installed. Their rating can be determined from

$$I_{\Delta n} = 50/Z_s$$

Where $I_{\Delta n}$ is the rated operating current of the device
50 is the touch voltage
Z_s is the measured loop impedence

RCD's can also be used for:

1 Discrimination e.g. a 100 mA device with time delay to protect the whole installation and a 30 mA for the sockets.
2 Protection against fire use, say, a 500 mA device.

All RCDs have a built-in test facility in the form of a test button. Operating this test facility creates an artificial out of balance condition that causes the device to trip. This only checks the mechanics of the tripping operation, it is not a substitute for the tests just discussed.

All other items of equipment such as switchgear, controlgear interlocks etc. must be checked to ensure that they are correctly mounted and adjusted and that they function correctly.

11
PERIODIC INSPECTION

Periodic inspection and testing

Circumstance which require a periodic inspection and test

Test and inspection is due; insurance, mortgage, licensing reasons; change of use; change of ownership; after additions or alterations; after damage; change of loading; to assess compliance with current regulations.

General reasons for a periodic inspection and test

1 To ensure the safety of persons and livestock.
2 To ensure protection of property from fire and heat.
3 To ensure that the installation is not damaged so as to impair safety.
4 To ensure that the installation is not defective and complies with the current regulations.

General areas of investigation

Safety; wear and tear; corrosion; damage; overloading; age; external influences; suitability; effectiveness.

Documentation to be completed

Periodic test report, schedule of test results and an inspection schedule.

Sequence of tests

1 Continuity of all protective conductors.
2 Polarity.
3 Earth fault loop impedance.
4 Insulation resistance.
5 Operation of isolating and switching devices.
6 Operation of RCDs.
7 Prospective fault current.

and where appropriate

8 Continuity of ring final circuit conductors.
9 Earth electrode resistance.
10 Manual operation of RCDs.
11 Protection by separation of circuits.
12 Insulation of non-conducting floors and walls.

This could be so simple. As it is, periodic inspection and testing tends to be complicated and frustrating. On the domestic scene, I doubt if any house owner actually decides to have a regular inspection. They say, 'If it works it must be OK'. It is usually only when there is a change of ownership that the mortgage companies insist on an electrical survey. The worst cases are, however, industry and commerce. Periodic inspections are requested, reluctantly, to satisfy insurers or an impending visit by the HSE. Even then it is usually the case that 'you can't turn that off' or 'why can't you just test this bit and then issue a certificate for the whole lot'. Under the rare circumstances when an inspection and test is genuinely requested it is difficult to convince the client that, as there are no drawings, or information about the installation, and that no switchgear is labelled etc., you are going to be on site for a considerable time and at a considerable cost.

When there are no drawings or items of information, especially on a large installation, there may be a degree of exploratory work to be carried out in order to ensure safety whilst inspecting and testing. If it is felt that it may be unsafe to continue with the inspection and test, then drawings and information **must** be produced in order to avoid contravening the Health and Safety at Work Act Section 6.

However, let us assume, as with the initial inspection, that the original installation was erected in accordance with the 16th edition, and that any alterations and/or additions have been faithfully recorded on the original documentation which is, of course, readily available!

A periodic inspection and test under these circumstances should be relatively easy, as little dismantling of the installation will be necessary, and the bulk of the work will be inspection.

Inspection should be carried out with the supply disconnected as it may be necessary to gain access to wiring in enclosures etc. and hence, with large installations it will probably need considerable liaison with the client to arrange convenient times for interruption of supplies to various parts of the installation.

This is also the case when testing protective conductors, as these must *never* be disconnected unless the supply can be isolated. This is particularly important for main equipotential bonding conductors which need to be disconnected in order to measure Z_e.

In general an inspection should reveal:

1 Any aspects of the installation that may impair the safety of persons and livestock against the effects of electric shock and burns.
2 That there are no installation defects that could give rise to heat and fire and hence damage property.
3 That the installation is not damaged or deteriorated so as to impair safety.
4 That any defects or non-compliance with the Regulations, that may give rise to danger, are identified.

As was mentioned earlier, dismantling should be kept to a minimum and hence a certain amount of sampling will take place.

This sampling would need to be increased in the event of defects being found.

From the testing point of view, not all of the tests carried out on the initial inspection may need to be applied. This decision depends on the condition of the installation.

The continuity of protective conductors is clearly important as is insulation resistance and loop impedance, but one wonders if polarity tests are necessary if the installation has remained undisturbed since the last inspection. The same applies to ring circuit continuity as the P–N test is applied to detect interconnections in the ring, which would not happen on their own.

It should be noted that if an installation is effectively supervised in normal use, then Periodic Inspection and Testing can be replaced by regular maintenance by skilled persons. This would only apply to, say, factory installations where there are permanent maintenance staff.

12
CERTIFICATION

Having completed all the inspection checks and carried out all the relevant tests, it remains to document all this information. This is done on electrical installation certificates, inspection schedules, test schedules, test result schedules, periodic inspection and test reports, minor works certificates and any other documentation you wish to append to the foregoing. Examples of such documentation are shown in the IEE Guidance Notes 3 on inspection and testing.

This documentation is vitally important. It has to be correct and signed or authenticated by a competent person. Electrical installation certificates and periodic reports must be accompanied by a schedule of test results and an inspection schedule for them to be valid. It should be noted that three signatures are required on an electrical installation certificate, one in respect of the design, one in respect of the construction and one in respect of the inspection and test. (For larger installations there may be more than one designer, hence the certificate has space for two signatures, i.e. designer 1 and designer 2.) It could be, of course, that for a very small company, one person signs all three parts. Whatever the case, the original must be given to the person ordering the work, and a duplicate retained by the contractor.

One important aspect of the electrical installation certificate is the recommended interval between inspections. This should be evaluated by the designer and will depend on the type of installation and its usage. In some cases the time interval is mandatory, especially where environments are subject to use by the public. Guidance Notes 3 give recommended maximum frequencies between inspections.

A periodic report form is very similar in part to an electrical installation certificate in respect of details of the installation, i.e. maximum demand, type of earthing system, Z_e etc. The rest of the form deals with the extent and limitations of the inspection and test, recommendations, and a summary of the installation. The record of the extent and limitations of the inspection is very important. It must be agreed with the client or other third party exactly what parts of the installation will be covered by the report and those that will not. The interval until the next test is determined by the inspector.

With regard to the schedule of test results, test values should be recorded unadjusted, any compensation for temperature etc. being made after the testing is completed.

Any alterations or additions to an installation will be subject to the issue of an electrical installation certificate, except where the addition is, say, a single point added to an existing circuit, then the work is subject to the issue of a minor works certificate.

Summarising:

(i) The addition of points to existing circuits require a Minor Works Certificate.
(ii) A new installation or an addition or alteration that comprises new circuits requires an Electrical Installation Certificate.
(iii) An existing installation requires a Periodic Test Report

Note: (ii) and (iii) must be accompanied by a schedule of test results and an inspection schedule.

Inspection & testing

As the client/customer is to receive the originals of any certification, it is important that *all* relevant details are completed correctly. This ensures that future inspectors are aware of the installation details and test results which may indicate a slow progressive deterioration in some or all of the installation.

These certificates etc. will also form part of a 'sellers pack' when a client wishes to sell a property.

The following is a general guide to completing the necessary documentation and should be read in conjunction with the examples given in BS 7671 and the On-site-guide.

Electrical installation certificate

1 **Details of client:**

Name: *Full name*
Address: *Full address and post code*
Description: *Domestic, industrial, commercial*
Extent: *What work has been carried out, e.g. full re-wire, new shower circuit etc. Tick a relevant box.*

2 **Designer/constructor/tester:**

Details of each or could be one person.
Note: Departures are not faults, they are systems/equipment etc. that are not detailed in BS 7671 but may be perfectly satisfactory.

3 **Next test:**

When the next test should be carried out and decided by the designer.

4 **Supply characteristics and earthing arrangements:**

Earthing system: *Tick relevant box (TT, TN-S etc.)*
Live conductors: *Tick relevant boxes*
Nominal voltage: *Obtain from supplier, but usually 230 V single phase U and U_0 but 400 V U and 230 U_0 for three phase*
Frequency: *From supplier but usually 50*
PFC: *From supplier or measured. Supplier usually gives 16 kA*
Z_e: *From supplier or measurement. Supplier usually gives 0.8 Ω for TN-S; 0.35 Ω for TN-C-S and 21 Ω for TT systems*

Main fuse: *Usually BS 1361, rating depends on maximum demand*

5 **Particulars of installation:**

Means of earthing: *Tick' suppliers facility' for TN systems 'earth electrode' for TT systems*

Maximum Demand: *Value without diversity*

Earth electrode: *Measured value or N/A*

Earthing and bonding:

Conductors: *Actual sizes and material, usually copper*

Main Switch or Circuit breaker (could be separate units or part of a consumer control unit): *BS number; Rating, current and voltage; Location; 'not address', i.e. where is it located in the building; Fuse rating if in a switch-fuse, else N/A; RCD details only if used as a main switch.*

6 **Comments on existing installation:**

Write down any defects found in other parts of the installation which may have been revealed during an addition or an alteration.

7 **Schedules:**

Indicate the number of test and inspection 'schedules that will accompany this certificate'

Periodic inspection report

1 **Details of client:**

Name: *Full name (could be a landlord etc.)*

Address: *Full address and post code (may be different to the installation address)*

Purpose: *E.g. Due date; change of owner/tenant; change of use etc.*

2 **Details of Installation:**

Occupier:	*Could be the client or a tenant*
Installation:	*Could be the whole or part (give details)*
Address:	*Full and post code*
Description:	*Tick relevant box*
Age:	*If not known, say so, or educated guess*
Alterations:	*Tick relevant box and insert age where known*
Last inspection:	*Insert date or 'not known'*
Records:	*Tick relevant box*

3 **Extent and limitations**
Full details of what is being tested (extent) and what is not (limitations) If not enough space on form add extra sheets.

4 **Next inspection**
Filled in by inspector and signed etc. under declaration.

5 **Supply details**
As per an Electrical Installation Certificate.

6 **Observations**
Tick relevant box, if work is required, record details and enter Relevant code (1, 2, 3 or 4) in space on right-hand side.

7 **Summary**
Comment on overall condition. Only Common sense and experience can determine whether satisfactory or unsatisfactory.

8 **Schedules**
Attach completed schedules of inspections and test results.

Minor electrical installation works certificate

Only to be used when simple additions or alterations are made, *not when a new circuit is added.*

1 **Description:**	*Full description of work*
Address:	*Full address*

Date: *Date when work was carried out*

Departures: *These are not faults, they are systems/ equipment etc. that are not detailed in BS 7671 but may be perfectly satisfactory (this is usually N/A)*

2 **Installation details**

Earthing: *Tick a relevant box*

Protection against Indirect contact: *99% of the time this will be EEBADS. Other methods should be recorded*

Protective device: *Enter type and rating. E.g. BS EN 60898 cb type B, 20 A*

Comments: *Note any defects/faults/omissions in other parts of the installation seen while conducting the minor works*

3 **Tests**

Earth continuity: *Measured and then tick in box if ok*

Insulation resistance: *Standard tests and results*

EFLI (Z_s): *Standard test and results*

Polarity: *Standard tests the tick in box if ok*

RCD: *Standard tests, record operating current and time*

4 **Declaration**

Name, address, signature etc.

Schedule of test results (as per BS 7671)

1 **Contractor**: *Full name of tester*

2 **Date:** *Date of test*

3 **Signature:** *Signature of tester*

4 **Method of protection against indirect contact:** *99% EEBADS but could be SELV etc.*

5 **Vulnerable equipment:** *Dimmers, electronic timers, CH controllers etc., i.e. anything electronic*

6 **Address:**	*Full, or if in a large installation, the location of a particular DB*
7 Earthing:	*Tick the relevant box*
8 Z_e at origin:	*Measured value*
9 PFC:	*Record the highest value i.e. PEFC or PSCC (Should be the same for TN-C-S)*
10 Instruments:	*Record serial numbers of each instrument, or one number for a composite instrument.*
11 Description:	*Suggest Initial or Periodic or whatever part of the installation is involved. E.g. Initial verification on a new shower circuit.*
12 kVA rating:	*Taken from the device. (Difficult when there are different devices in an installation) Nothing to stop adding sheets to this form!*
13 Type and rating:	*E.g. BS EN 60898 cb type B, 32A, or BS 88 40A etc.*
14 Wiring conductors:	*Size of Live and CPC, e.g.* $2.5\,mm^2/1.5\,mm^2$
15 Test results:	*Fill in all measured values (R1 + R2) etc. Tick box if ring P-N is ok. If any test does not appear on the sheet, e.g.* $5 \times I_{\Delta n}$*, write the results in the remarks column.*

Schedule of inspections (as per BS 7671)

Do not leave boxes uncompleted

N/A in a box if it is not relevant

✔ in a box if has been inspected and is ok

X in a box if it has been inspected and is incorrect.

APPENDIX A SAMPLE PAPER

Section A short answer

1 Indicate three main areas, about which you would require information, in order correctly to carry out an initial verification of a new installation.

2 There are various documents that are relevant to the inspection and testing of an installation, state:

(a) one statutory item of documentation
(b) two non-statutory items of documentation.

3 An Electrical Installation Certificate should be accompanied by signed documentation regarding three stages of an installation. What are these stages?

4 Apart from wear and tear state three areas of investigation that you would consider when carrying out a periodic inspection and test of an installation.

5 State three human senses that could be used during an inspection of an installation.

6 During a test on an installation, the following readings were obtained:

20 MΩ; 8 kA; 22 ms

List the instruments which gave these readings.

7 The following circuits are to be tested for insulation resistance. State the test voltages to be applied and the minimum acceptable value of insulation resistance in each case:

(a) SELV circuit
(b) LV circuit up to 500 V
(c) LV circuit over 500 V.

8 List the first three tests that should be carried out during an initial verification on a new domestic installation.

9 The test for the continuity of a cpc in a radial circuit feeding one socket outlet, uses a temporary link and a milli-ohmmeter, state:

(a) where the temporary link is connected
(b) where the milli-ohmmeter is connected
(c) what does the meter reading represent?

10 List three different protective conductors that would need to be connected to the main earthing terminal of an installation.

11 The following readings were obtained during the initial tests on a healthy ring final circuit:

P1–P2–0.8 Ω; N1–N2–0.8 Ω;
cpc1–cpc2–0.8 Ω

(a) what readings would you expect:
  (i) between P and N conductors at each socket outlet?
  (ii) between P and cpc at each socket outlet?
(b) what the P to cpc reading represents?

12 What happens to:

(a) conductor resistance when conductor length increases?
(b) insulation resistance when cable length increases?
(c) conductor resistance when conductor area increases?

13 List three precautions to be taken prior to commencing an insulation resistance test on an installation.

14 An enclosure has been fabricated on site to house electrical equipment. State the IP codes that the enclosure should at least comply with.

15 What degree of protection is offered by enclosures offering the following:

(a) IP XXB
(b) IP4X
(c) IPX8.

16 List three reasons for conducting a dead polarity test on an installation.

17 What earthing systems are attributed to the following:

(a) an overhead line supply with no earth?
(b) a multicore supply cable with a separate neutral and earth?
(c) a supply cable in which the functions of earth and neutral are performed by one conductor?

18 State three locations where special considerations should be made with regards to electrical installations.

19 From the formula

$$Z_s = Z_e + \frac{(R_1 + R_2) \times 1.2 \times L}{1000}$$

what is represented by:

(a) Z_e?
(b) R_2?
(c) 1.2?

20 State any three functional tests that may be carried out on a domestic installation.

Section B

Figure 12.1 shows the layout of the electrical installation in a new detached garage. You are to carry out an initial verification of that installation.

1 (a) What documentation/information will you require in order to carry out the verification?
  (b) Where should it be located?
  (c) What particularly important details regarding this installation should have been included on such documentation?
  (d) What consideration should be given to the existing installation from which this new installation is fed?

2 List five areas of inspection for this installation that should be carried out prior to testing.

3 The following test results were obtained from a ring final circuit continuity test. State if the readings for each socket are

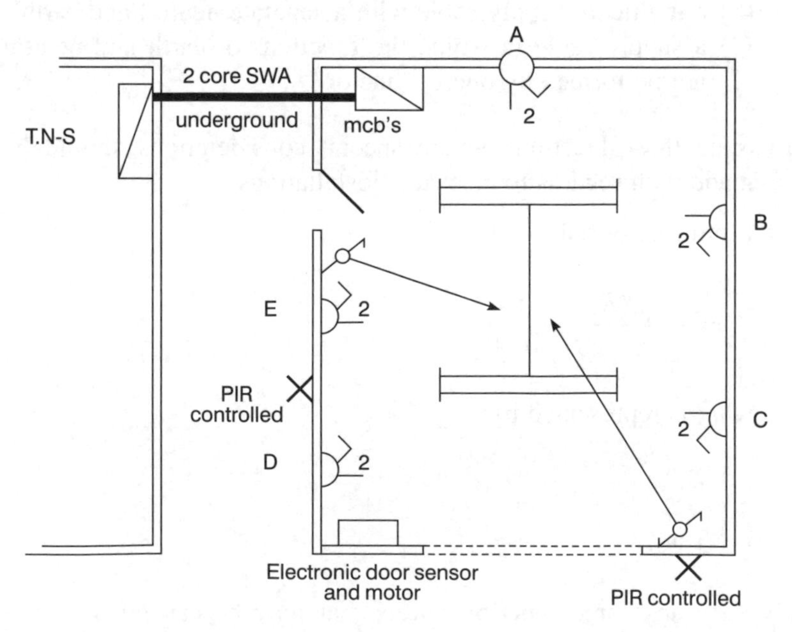

Figure 12.1

satisfactory and give reasons for those readings you feel are unsatisfactory.

Phase, neutral and cpc loops = 0.5 Ω

Socket	*P–N*	*P–cpc*
A	0.25	0.26
B	No reading	0.25
C	0.35	0.24
D	0.24	No reading
E	0.26	0.26

4 (a) Describe in detail how you would carry out an insulation resistance test on this installation.
(b) The test result indicates an overall value of 1.75 MΩ, what actions, if any, should be taken. Explain your reasons.

5 A loop impedance test on the lighting circuit cannot be conducted, as the 6A Type B MCB keeps tripping out. Explain why this is, and how the problem may be overcome in order to conduct the test.

6 (a) The electronic door sensor/motor is wired on its own radial circuit, list all the component parts of the earth fault loop path associated with this circuit in the event of a fault to earth.
(b) If the maximum value of loop impedance for this circuit is 2.4 Ω and an earth fault causes a current of 120 A, show by calculation if this value will disconnect the circuit in the required time.

APPENDIX B SUGGESTED SOLUTIONS TO SAMPLE PAPER

Section A

1 Any three of:

- to ensure accessories etc. to relevant standard
- to ensure compliance with BS 7671
- to ensure no damage that may cause danger.

(3 marks)

2 (a) *The Electricity at Work Regulations* **(1 mark)**

(b) any two of:

- BS 7671
- *Guidance Note 3*
- *The On-site Guide*.

(2 marks)

3
- design
- construction
- inspection and testing.

(3 marks)

4 Any three of:

- to ensure safety of persons and livestock
- to ensure protection from fire and heat

- to ensure that the installation is not damaged so as to impair safety
- to ensure that the installation is not defective and complies with current regulations.

(3 marks)

5
- visual
- touch
- smell.

(3 marks)

6
- insulation resistance tester
- prospective short circuit current tester
- RCD tester.

(3 marks)

7
- 250 V and 0.25 MΩ
- 500 V and 0.5 MΩ
- 1000 V and 1.0 MΩ.

(3 marks)

8
- continuity of protective conductors
- continuity of ring final circuit conductors
- insulation resistance.

(3 marks)

9
- between P and E at the consumer unit
- between P and E at the socket outlet
- this value is $(R_1 + R_2)$ for the circuit.

(3 marks)

10 Any three of:
- circuit protective conductor
- main equipotential bonding conductor
- earthing conductor
- lightning conductor.

(3 marks)

11
- 0.4 Ω
- 0.4 Ω
- $(R_1 + R_2)$ for the ring.

(3 marks)

12
- increases
- decreases
- decreases.

(3 marks)

13 Any three of:

- check on existence of electronic equipment
- check there are no neons, capacitors etc. in circuit
- all switches closed and accessories equipment removed
- no danger to persons or livestock by conducting the test.

(3 marks)

14
- IP 2X
- IP XXB
- IP 4X.

(3 marks)

15
- finger contact only
- small foreign solid bodies or 1 mm diameter wires
- total submersion.

(3 marks)

16
- all single pole devices in phase conductor only
- centre contact of Edison screw lampholders in phase conductor
- all accessories correctly connected.

(3 marks)

17
- TT
- T–N–S
- TN–C–S.

(3 marks)

18 Any three of:

- bathrooms and shower basins
- hot air saunas
- swimming pools
- construction sites
- agricultural and horticultural situations
- restrictive conductive locations
- high earth leakage locations
- caravans
- caravan parks
- highway power supplies and street furniture.

(3 marks)

19 • external loop impedance
• resistance of cpc
• multiplier for conductor operating temperature.

(3 marks)

20 Any three of:
• test button operation of an RCD
• operation of dimmer switch
• operation of main isolating switch
• operation of MCBs
• operation of two-way switching.

(3 marks)

Section B

1 (a) The results of the assessment of general characteristics section 311, 312 and 313, and diagrams charts and similar information regarding the installation

(5 marks)

(b) In or adjacent to the distribution board **(3 marks)**

(c) Reference to the electronic door sensor and the PIR controlled external luminaires as these could be vulnerable to a typical test **(3 marks)**

(d) Maximum demand, rating of consumer unit, earthing and bonding arrangements, capacity of main protective device etc. **(4 marks)**

2 Any five relevant areas from the check list. **(3 marks each)**

Items inspected

☐ 1 Connection of conductors

☐ 2 Identification of conductors

☐ 3 Routing of cables in safe zones or protected against mechanical damage

☐ 4 Selection of conductors for current and voltage drop

☐ 5 Connection of single-pole devices for protection or switching in phase conductors only

☐ 6 Correct connection of socket-outlets and lampholders

☐ 7 Presence of fire barriers and protection against thermal effects

8 Method of protection against electric shock

(a) Protection against both direct and indirect contact

☐ SELV

☐ Limitation of discharge of energy

(b) Protection against direct contact

☐ Insulation of live parts

☐ Barrier or enclosure

☐ Obstacles

☐ Placing out of reach

☐ PELV

(c) Protection against indirect contact

(i) Earthed equipotential bonding and automatic disconnection of supply

☐ Presence of earthing conductors

☐ Presence of protective conductors

☐ Presence of main equipotential bonding conductors

☐ Presence of supplementary equipotential bonding conductors

(ii) Use of Class II equipment or equivalent insulation

(iii) Non-conducting location

☐ Absence of protective conductors

(iv) Earth-free local equipotential bonding

☐ Presence of earth-free equipotential bonding conductors

(v) Electrical separation (413–06)

☐ 9 Prevention of mutual detrimental influence

☐ 10 Presence of appropriate devices for isolating and switching correctly located

☐ 11 Presence of undervoltage protective devices where appropriate

12 Choice and setting of protective and monitoring devices (for protection against indirect contact and/or overcurrent)

☐ Residual current devices

☐ Overcurrent devices

☐ 13 Labelling of protective devices, switches and terminals

☐ 14 Selection of equipment and protective measures appropriate to external influences

☐ 15 Adequacy of access to switchgear and equipment

☐ 16 Presence of danger notices and other warnings

☐ 17 Presence of diagrams, instructions and necessary information

☐ 18 Erection methods

☐ 19 Requirements of special locations

Tick to indicate item has been inspected
Delete if item not applicable

3 • Socket A OK as readings are approximately $\frac{1}{2}$ of 0.5 Ω **(3 marks)**

- Socket B cross-polarity P–cpc, or twisted N conductors not in N terminal **(3 marks)**
- Socket C loose neutral connection **(3 marks)**
- Socket D cross-polarity P–N, or twisted CPCs not in terminal **(3 marks)**
- Socket E OK. **(3 marks)**

4 (a) Conduct the test from the house as this will then include the SWA cable. Disconnect the supply to the PIR controlled lights and the electronic door sensor. Disconnect the capacitor and

ballast at each fluorescent luminaire. With the garage main switch and the MCBs ON and any accessories unplugged, test at 500 V between live conductors connected together and earth and then between each live conductor. Operate the two-way switches during each test. The test readings should not be less than 0.5 MΩ.

(8 marks)

(b) If any reading is below 2 MΩ, then there may be a latent defect and then each circuit should be tested separately and the insulation resistance in each case should be greater than 2 MΩ. **(7 marks)**

5 As a loop impedance tester delivers a high current for a short time, it is not unusual for sensitive MCBs with low ratings to trip out on overload. The loop impedance in such cases will have to be determined by a combination of measurement and calculation as follows:

Measure Z_e and measure $(R_1 + R_2)$ for the circuit, then $Z_s = Z_c + (R_1 + R_2)$ **(7 marks)**

6 (a)

- The point of fault
- The cpc
- The steel wire armour of the garage supply
- The earthing conductor
- The metallic earth return path of the supply cable
- The earthed neutral of the transformer
- The transformer winding
- Phase conductors.

(8 marks)

(b) $Z_s = U_{oc}/I$

$= 240/120$

$= 2\,\Omega$

so circuit protection will operate fast enough.

(7 marks)

APPENDIX C THE 2004 VERSION OF BS 7671: 2001 (INCORPORATING AMENDMENTS 1:2002 & 2:2004) — THE CHANGES

1 The Electricity Supply Regulations 1988 have now been replaced by 'The Electricity Safety Quality and Continuity Regulations 2002'.
2 New conductor/cable colours (this appendix deals only with AC circuits).

From the 1st April 2004 new colours were introduced. Old *or* new colours may be used for new installations until 1st April 2006 after which only the *new* colours are acceptable. CPC colours remain the same.

These new colours are as follows:

SINGLE PHASE:

Phase: Brown
Neutral: Blue

THREE PHASE:

Phase: Brown or Black or Grey (if all brown or all black or all grey are used, they must still be identified brown, black & grey at terminations)
Numbers and letters may also be used to identify phases using L1, L2 and L3
Neutral: Blue

For a colour illustration of the interface between old and new please see the inside back cover.

INDEX

Index